GENERAL INFORMATION CONCERNING PATENTS

U.S. DEPARTMENT OF COMMERCE
William M. Daley, Secretary

PATENT AND TRADEMARK OFFICE
Bruce A. Lehman, Assistant Secretary and
Commissioner of Patents and Trademarks

Reprinted August 1997

Contents

Functions of the Patent and Trademark Office	5
Purpose of this booklet	6
What is a patent?	7
Patent laws	9
What can be patented	10
Novelty and other conditions for obtaining a patent	12
The United States Patent and Trademark Office	13
Publications of the Patent and Trademark Office	15
General information and correspondence	19
Library, search room services and Patent and Trademark Depository Libraries	21
Attorneys and agents	28
Disclosure Document	31
Who may apply for a patent	33
Application for patent	34
Oath or declaration, signature	37
Filing Fees	38
Specification (description and claims)	40
Drawing	43
Models, exhibits, specimens	54
Examination of applications and proceedings in the Patent and Trademark Office	55
Amendments to application	58
Time for response and abandonment	61
Appeal to the Board of Patent Appeals and Interferences and to the courts	62
Interferences	63
Allowances and issue of patent	65
Nature of patent and patent rights	66
Maintenance Fees	68
Correction of patents	69

Assignments and licenses	70
Infringement of patents	73
Patent marking and "patent pending"	74
Design patents	75
Plant patents	76
Treaties and foreign patents	79
Foreign applicants for United States patents	82
Answers to questions frequently asked	84

Functions of the Patent and Trademark Office

The Patent and Trademark Office (PTO) is an agency of the U.S. Department of Commerce. The role of the Patent and Trademark Office is to grant patents for the protection of inventions and to register trademarks. It serves the interest of inventors and businesses with respect to their inventions and corporate products, and service identifications. It also advises and assists the bureaus and offices of the Department of Commerce and other agencies of the Government in matters involving “intellectual property” such as patents, trademarks and semiconductor mask works. Through the preservation, classification, and dissemination of patent information, the Office aids and encourages innovation and the scientific and technical advancement of the nation.

In discharging its patent related duties, the Patent and Trademark Office examines applications and grants patents on inventions when applicants are entitled to them; it publishes and disseminates patent information, records assignments of patents, maintains search files of U.S. and foreign patents, and maintains a search room for public use in examining issued patents and records. It supplies copies of patents and official records to the public. Similar functions are performed relating to trademarks.

Purpose of this Booklet

The purpose of this booklet is to give users some general information about patents and the operations of the Patent and Trademark Office. It attempts to answer many of the questions commonly asked of the Patent and Trademark Office but is not intended to be a comprehensive textbook on patent law or a guide for the patent attorney. It is hoped that this information will be useful to inventors and prospective applicants for patents, to students, and to others who may be interested in patents by giving them a brief general introduction to the subject.

Additional information may be obtained from the publications listed under the sections "Publications of the Patent and Trademark Office" and "General information and correspondence." Also, information is available on the U.S. Patent and Trademark Office's Web site at: *http//www.uspto.gov.* The Patent and Trademark Office does not publish any textbooks on patent law, but a number of such works for the specialist and for the general reader have been published by private concerns.

What Is a Patent?

A patent for an invention is a grant of a property right by the Government to the inventor (or his or her heirs or assigns), acting through the Patent and Trademark Office. The term of the patent shall be 20 years from the date on which the application for the patent was filed in the United States or, if the application contains a specific reference to an earlier filed application under 35 U.S.C. 120, 121 or 365(c), from the date the earliest such application was filed, subject to the payment of maintenance fees. The right conferred by the patent grant extends only throughout the United States and its territories and possessions.

The right conferred by the patent grant is, in the language of the statute and of the grant itself, "the right to exclude others from making, using, offering for sale, or selling" the invention in the United States or "importing" the invention into the United States. What is granted is not the right to make, use, offer for sale, sell or import, but the right to exclude others from making, using, offering for sale, selling or importing the invention. Most of the statements in the preceding paragraphs will be explained in greater detail in later sections.

Some persons occasionally confuse patents, copyrights, and trademarks. Although there may be some resemblance in the rights of these three kinds of intellectual property, they are different and serve different purposes.

Copyrights

A copyright protects the writings of an author against copying. Literary, dramatic, musical and artistic works are included within the protection of the copyright law, which in some instances also confers performing and recording rights. The copyright protects the form of expression rather than to the subject matter of the writing. A description of a machine could be copyrighted as a writing, but this would only prevent others from copying the description; it would not prevent others from writing a description of their own or from making and using the machine. Copyrights are registered in the Copyright Office in the Library of Congress. Information concerning copyrights may be obtained from the Register of Copyrights, Library of Congress, Washington, D.C. 20559. (Telephone 202-707-3000)

Trademarks/Servicemarks

A trademark or servicemark relates to any word, name, symbol or device which is used in trade with goods or services to indicate the source or origin of the goods or services and to distinguish them from the goods or services of others. Trademark rights may be used to prevent others from using a confusingly similar mark but not to prevent others from making the same goods or from selling them under a non-confusing mark. Similar rights may be acquired in marks used in the sale or advertising of services (service marks). Trademarks and service marks which are used in interstate or foreign commerce may be registered in the Patent and Trademark Office. The procedure relating to the registration of trademarks and some general information concerning trademarks is given in a separate pamphlet entitled "Basic Facts About Trademarks."

Patent Laws

The Constitution of the United States gives Congress the power to enact laws relating to patents, in Article I, section 8, which reads "Congress shall have power . . . to promote the progress of science and useful arts, by securing for limited times to authors and inventors the exclusive right to their respective writings and discoveries." Under this power Congress has from time to time enacted various laws relating to patents. The first patent law was enacted in 1790. The law now in effect is a general revision which was enacted July 19, 1952, and which came into effect January 1, 1953. It is codified in Title 35, United States Code.

The patent law specifies the subject matter for which a patent may be obtained and the conditions for patentability. The law establishes the Patent and Trademark Office to administer the law relating to the granting of patents, and contains various other provisions relating to patents.

What Can Be Patented

The patent law specifies the general field of subject matter that can be patented and the conditions under which a patent may be obtained.

In the language of the statute, any person who "invents or discovers any new and useful process, machine, manufacture, or composition of matter, or any new and useful improvement thereof, may obtain a patent," subject to the conditions and requirements of the law. The word "process" is defined by law as a process, act or method, and primarily includes industrial or technical processes. The term "machine" used in the statute needs no explanation. The term "manufacture" refers to articles which are made, and includes all manufactured articles. The term "composition of matter" relates to chemical compositions and may include mixtures of ingredients as well as new chemical compounds. These classes of subject matter taken together include practically everything which is made by man and the processes for making the products.

The Atomic Energy Act of 1954 excludes the patenting of inventions useful solely in the utilization of special nuclear material or atomic energy for atomic weapons.

The patent law specifies that the subject matter must be "useful." The term "useful" in this connection refers to the condition that the subject matter has a useful purpose and also includes operativeness, that is, a machine which will not operate to perform the intended purpose would not be called useful, and therefore would not be granted a patent.

Interpretations of the statute by the courts have defined the limits of the field of subject matter which can be patented, thus it has been held that the laws of nature, physical phenomena and abstract ideas are not patentable subject matter.

A patent cannot be obtained upon a mere idea or suggestion. The patent is granted upon the new machine, manufacture, etc., as has been said, and not upon the idea or suggestion of the new machine. A complete description of the actual machine or other subject matter for which a patent is sought is required.

Novelty and Other Conditions for Obtaining a Patent

In order for an invention to be patentable it must be new as defined in the patent law, which provides that an invention cannot be patented if:

"(a) the invention was known or used by others in this country, or patented or described in a printed publication in this or a foreign country, before the invention thereof by the applicant for patent," or "(b) the invention was patented or described in a printed publication in this or a foreign country or in public use or on sale in this country more than one year prior to the application for patent in the United States . . ."

If the invention has been described in a printed publication anywhere in the world, or if it has been in public use or on sale in this country before the date that the applicant made his/her invention, a patent cannot be obtained. If the invention has been described in a printed publication anywhere, or has been in public use or on sale in this country more than one year before the date on which an application for patent is filed in this country, a patent cannot be obtained. In this connection it is immaterial when the invention was made, or whether the printed publication or public use was by the inventor himself/herself or by someone else. If the inventor describes the invention in a printed publication or uses the invention publicly, or places it on sale, he/she must apply for a patent before one year has gone by, otherwise any right to a patent will be lost.

Even if the subject matter sought to be patented is not exactly shown by the prior art, and involves one or more differences over the most nearly similar thing already known, a patent may still be refused if the differences would be obvious. The subject matter sought to be patented must be sufficiently different from what has been used or described before that it may be said to be nonobvious to a person having ordinary skill in the area of technology related to the invention. For example, the substitution of one material for another, or changes in size, are ordinarily not patentable.

The United States Patent and Trademark Office

Congress established the United States Patent and Trademark Office to issue patents on behalf of the Government. The Patent and Trademark Office as a distinct bureau may be said to date from the year 1802 when a separate official in the Department of State who became known as "Superintendent of Patents" was placed in charge of patents. The revision of the patent laws enacted in 1836 reorganized the Patent and Trademark Office and designated the official in charge as Commissioner of Patents and Trademarks. The Patent and Trademark Office remained in the Department of State until 1849 when it was transferred to the Department of Interior. In 1925 it was transferred to the Department of Commerce where it is today.

The Patent and Trademark Office administers the patent laws as they relate to the granting of patents for inventions, and performs other duties relating to patents. It examines applications for patents to determine if the applicants are entitled to patents under the law and grants the patents when they are so entitled; it publishes issued patents and various publications concerning patents, records assignments of patents, maintains a search room for the use of the public to examine issued patents and records, supplies copies of records and other papers, and the like. Similar functions are performed with respect to the registration of trademarks. The Patent and Trademark Office has no jurisdiction over questions of infringement and the enforcement of patents, nor over matters relating to the promotion or utilization of patents or inventions.

The head of the Office is the Assistant Secretary of Commerce and Commissioner of Patents and Trademarks, and his staff includes the Deputy Assistant Secretary of Commerce and Deputy Commissioner of Patents and Trademarks, several assistant commissioners, and other officials. As head of the Office, the Commissioner superintends or performs all duties respecting the granting and issuing of patents and the registration of trademarks; exercises general supervision over the entire work of the Patent and Trademark Office; prescribes the rules, subject to the approval of the Secretary of Commerce, for the conduct of proceedings in the Patent and Trademark Office, and for recognition of attorneys and agents; decides various questions brought before him by petition as prescribed by the rules; and performs other duties necessary and required for the administration of the Patent and Trademark Office.

The work of examining applications for patents is divided among a number of examining groups, each group having jurisdiction over certain assigned fields of technology. Each group is headed by a group director and staffed by examiners. The examiners review applications for patents and determine whether patents can be granted. An appeal can be taken to the Board of Patent Appeals and Interferences from their decisions refusing to grant a patent, and a review by the Commissioner of Patents and Trademarks may be had on other matters by petition. The examiners also identify applications that claim the same invention and initiate proceedings, known as interferences, to determine who was the first inventor.

In addition to the examining groups, other offices perform various services, such as receiving and distributing mail, receiving new applications, handling sales of printed copies of patents, making copies of records, inspecting drawings, and recording assignments.

At present, the Patent and Trademark Office has about 5,700 employees, of whom about half are examiners and others with technical and legal training. Patent applications are received at the rate of over 200,000 per year. The Patent and Trademark Office receives over five million pieces of mail each year.

Publications of the Patent and Trademark Office

U.S. Patents. —The specification and accompanying drawings of all patents are published on the day they are granted and printed copies are sold to the public by the Patent and Trademark Office. Over 6,000,000 patents have been issued.

Printed copies of any patent, identified by its patent number, may be purchased from the Patent and Trademark Office. Current fee schedule is available by calling the PTO General Information Services at 1-800-786-9199 or 703-308-4357 or by accessing PTO's Web site at *http://www.uspto.gov.*

Future patents classified in subclasses containing subject matter of interest may be obtained, as they issue, by prepayment of a deposit and a service charge. For the cost of such subscription service, a separate inquiry should be sent to the Patent and Trademark Office.

Official Gazette of the United States Patent and Trademark Office. — The Official Gazette of the United States Patent and Trademark Office is the official journal relating to patents and trademarks. It has been published weekly since January 1872 (replacing the old "Patent Office Reports"), and is now issued each Tuesday in two parts, one describing patents and the other trademarks. It contains a claim and a selected figure of the drawings of each patent granted on that day; notices of patent and trademark lawsuits; indexes of patents and patentees; list of patents available for license or sale; a list of Patent and Trademark Depository Libraries (PTDLs); and much general information such as orders, notices, changes in rules, changes in classification, etc. The Official Gazette is sold on subscription and by single copies by the Superintendent of Documents, U.S. Government Printing Office, Washington, D.C. 20402.

The illustrations and claims of the patents are arranged in the Official Gazette according to the Patent and Trademark Office classification of subject matter, permitting ready reference to patents in any particular technology. Copies of the Official Gazette are available in the PTDLs and public libraries of other cities.

Index of Patents. —A two-part publication which summarizes for a given calendar year the classification and inventor/assignee information at the time of issue for utility, Design, Reissue, and Plant Patents; Reexamination Certificates, and Statutory Invention Registrations published weekly in the Official Gazette. Part I, List of Patentees comprises an alphabetical listing of every patentee and assignee recorded at the time the patent document was issued. Part II, Index to Subjects of Inventions comprises a listing of all patents for the year according to U.S. Patent Classification class and subclass designation at the time the patent document was issued—it is not an index of subjects, per se. Sold by the Superintendent of Documents.

Index of Trademarks. —An annual index of registrants of trademarks. Sold by the Superintendent of Documents.

Index to the U.S. Patent Classification System. —An alphabetical list of approximately 65,000 common, informal headings or terms which refer to specific classes and subclasses in the Manual of Classification used to categorize patents. It is intended as a means for initial entry into the Classification System and should be especially useful for persons not familiar with the system as well as those who may not be familiar with any particular technology under study. Sold by the Superintendent of Documents.

Manual of Classification. —A loose-leaf manual containing a list of all the classes and subclasses of inventions in the Patent and Trademark Office classification systems, a subject matter index, and other information relating to U.S. patent classification system. Each subclass has a short, descriptive title often arranged in a specific hierarchical order designated by dots for indentation levels. Substitute pages are issued from time to time. Annual subscription includes the basic manual and substitute pages. Sold by the Superintendent of Documents.

Classification Definitions. —Gives a detailed definition for each class and official subclass included in the Manual of Classification. The definitions indicate the subject matter to be found in or excluded from a class or subclass; they limit or expand in precise manner the meaning intended for each subclass title; they serve as a guide to users of the Manual of Classification to refer to the same subclass for patents on a particular technology by eliminating, as much as possible, subjective and varying interpretations of the meanings of subclass titles. The "notes" illustrate the kinds of information that can be found in a subclass and direct the searcher to other related subclasses which may contain relevant information. Subscription service consists of a basic full set of definitions and semiannual sets of updated definitions for an indeterminate period. Sold by the Superintendent of Documents.

Title 37 Code of Federal Regulations. —Includes rules of practice for Patents, Trademarks, and Copyrights. Available from the Superintendent of Documents.

Basic Facts about Trademarks. —Contains general information for the layman about applications for, and registration of, trademarks and service marks. Sold by the Superintendent of Documents.

Attorneys and Agents Registered to Practice Before the U.S. Patent and Trademark Office. —An alphabetical and geographical listing of patent attorneys and agents registered to practice before the U.S. Patent and Trademark Office. Sold by the Superintendent of Documents.

Manual of Patent Examining Procedure (MPEP). —A loose-leaf manual which serves primarily as a detailed reference work on patent examining practice and procedure for the Patent and Trademark Office's Examining Corps. Subscription service includes basic manual, periodic revisions, and change notices. Sold by the Superintendent of Documents. (The MPEP is also available in electronic form from the PTO's Office of Electronic Information and as an Internet information file.)

Guide for the Preparation of Patent Drawings. —A collection of the most pertinent rules from Title 37 of the Code of Federal Regulations pertaining to patent drawings with interpretations of those rules and examples. Sold by the Superintendent of Documents.

PTO Products and Services Catalog, Information Dissemination Organizations (IDO). —This compendium describes the products and services available from the Patent and Trademark Office (PTO) and provides ordering information. Many items are available free of charge from the Information Dissemination Organizations (IDOs). The catalog lists products and services available from the three IDO offices responsible for public records; electronic products/services; general information; PTO's public search facilities in Arlington, Virginia; and the network of 80 Patent and Trademark Depository Libraries (PTDLs) in 49 states, the District of Columbia, and Puerto Rico. Included is detailed descriptions of data available on magnetic tape, diskette, or CD-ROM. A number of products can be ordered in paper form, including PTO forms and patent or trademark copies, as well as Technology Assessment and Forecast (TAF) statistical reports and research publications. Available from the PTO General Information Services at 1-800-786-9199 or 703-308-4357.

Many publications listed above, as well as other Patent and Trademark Office products and services are available electronically from IDO or on the PTO's Web site. Call General Information Services for more information at 800-786-9199 or 703-308-4357.

General Information and Correspondence

All business with the Patent and Trademark Office should be transacted in writing and all correspondence relating to patent matters should be addressed to "ASSISTANT COMMISSIONER FOR PATENTS, WASHINGTON, D.C. 20231." Correspondents should be sure to include their full return addresses, including zip codes. The principal location of the PTO is Crystal Plaza 3, 2021 Jefferson Davis Highway, Arlington, Virginia. The personal presence of applicants at the PTO is unnecessary.

Applicants and attorneys are required to conduct their business with decorum and courtesy. Papers presented in violation of this requirement will be returned.

Separate letters (but not necessarily in separate envelopes) should be written for each distinct subject of inquiry, such as assignments, payments, orders for printed copies of patents, orders for copies of records, and requests for other services. None of these inquiries should be included with letters responding to Office actions in applications.

When a letter concerns a patent application, the correspondent must include the application number, filing date, and Group Art Unit number. When a letter concerns a patent, it must include the name of the patentee, the title of the invention, the patent number, and the date of issue.

An order for a copy of an assignment must give the book and page, or reel and frame of the record, as well as the name of the inventor; otherwise, an additional charge is made for the time consumed in making the search for the assignment.

Applications for patents are not open to the public, and no information concerning them is released except on written authority of the applicant, his/her assignee, or his/her attorney, or when necessary to the conduct of the business of the PTO. Patents and related records, including records of any decisions, the records of assignments other than those relating to assignments of patent applications, books, and other records and papers in the Office are open to the public. They may be inspected in the Patent and Trademark Office Search Room or copies may be ordered.

The Office cannot respond to inquiries concerning the novelty and patentability of an invention in advance of the filing of an application; give advice as to possible infringement of a patent; advise of the propriety of filing an application; respond to inquiries as to whether, or to whom, any alleged invention has been patented; act as an expounder of the patent law or as counselor for individuals, except in deciding questions arising before it in regularly filed cases. Information of a general nature may be furnished either directly or by supplying or calling attention to an appropriate publication.

Library, Search Room Services and Patent and Trademark Depository Libraries

The Scientific and Technical Information Center of the Patent and Trademark Office located at Crystal Plaza 3, 2C01, 2021 Jefferson Davis Highway, Arlington, VA, has available for public use over 120,000 volumes of scientific and technical books in various languages, about 90,000 bound volumes of periodicals devoted to science and technology, the official journals of 77 foreign patent organizations, and over 40 million foreign patents on paper, microfilm, microfiche, and CD-ROM. The Scientific and Technical Information Center is open to the public from 8:30 a.m. to 5:00 p.m., Monday through Friday except federal holidays.

The Patent Search Room located at Crystal Plaza 3, 1A01, 2021 Jefferson Davis Highway, Arlington, VA, is provided where the public may search and examine United States patents granted since 1790. Patents are arranged according to the U.S. Patent Classification System of over 400 classes and over 136,000 subclasses. By searching in these classified groupings of patents, it is possible to determine, before actually filing an application, whether an invention has been anticipated by a United States patent, and it is also possible to obtain the information contained in patents relating to any field of endeavor. The Patent Search Room contains a set of United States patents arranged in numerical order and a complete set of the Official Gazette.

A Files Information Room also is maintained where the public may inspect the records and files of issued patents and other open records. Applicants, their attorneys or agents, and the general public are not entitled to use the records and files in the examiners' rooms.

The Patent Search Room is open from 8 a.m. to 8 p.m. Monday through Friday except on Federal holidays.

Since a patent is not always granted when an application is filed, many inventors attempt to make their own search of the prior patents and publications before applying for a patent. This may be done in the Patent Search Room of the Patent and Trademark Office, and in libraries, located throughout the U.S., which have been designated as Patent and Trademark Depository Libraries (PTDLs). An inventor may make a preliminary search through the United States patents to discover if the particular invention or one similar to it has been shown in the prior patent. An inventor may also employ patent attorneys or agents to perform the preliminary search. This search is not always as complete as that made by the Patent and Trademark Office during the examination of an application, but only serves, as its name indicates, a preliminary purpose. For this reason, the Patent and Trademark Office examiner may, and often does, reject claims in an application on the basis of prior patents or publications not found in the preliminary search.

Those who cannot come to the Patent Search Room may order from the Patent and Trademark Office copies of lists of original patents or of cross-referenced patents contained in the subclasses comprising the field of search, or may inspect and obtain copies of the patents at a Patent and Trademark Depository Library. The Patent and Trademark Depository Libraries (PTDLs) receive current issues of U.S. Patents and maintain collections of earlier issued patents and trademark information. The scope of these collections varies from library to library, ranging from patents of only recent years to all or most of the patents issued since 1790.

These patent collections are open to public use. Each of the PTDLs, in addition, offers the publications of the U.S. Patent Classification System (e.g., Manual of Classification, Index to the U.S. Patent Classification System, Classification Definitions, etc.) and other patent documents and forms, and provides technical staff assistance in their use to aid the public in gaining effective access to information contained in patents. The collections are organized in patent number sequence.

Available in all PTDLs is the Cassis CD-ROM system. With various files, it permits the effective identification of appropriate classifications to search, provides numbers of patents assigned to a classification to facilitate finding the patents in a numerical file of patents, provides the current classification(s) of all patents, permits word searching on classification titles, and on abstracts, and provides certain bibliographic information on more recently issued patents.

Facilities for making paper copies from microfilm, the paper bound volumes or CD-ROM are generally provided for a fee.

Due to variations in the scope of patent collections among the PTDLs and in their hours of service to the public, anyone contemplating the use of the patents at a particular library is advised to contact that library, in advance, about its collection, services, and hours, so as to avert possible inconvenience.

State		Name of Library	Telephone #
Alabama	*	Auburn: Ralph Brown Draughon Library, Auburn University	334-844-1747
		Birmingham Public Library	205-226-3620
Alaska		Anchorage: Z. J. Loussac Public Library, Anchorage Municipal Libraries	907-562-7323
Arizona	*	Tempe: Noble Science and Engineering Library, Arizona State University	602-965-7010
Arkansas	*	Little Rock: Arkansas State Library	501-682-2053
California	*	Los Angeles Public Library	213-228-7220
		Sacramento: California State Library	916-654-0069
		San Diego Public Library	619-236-5813
	*	San Francisco Public Library	415-557-4500
	**	Sunnyvale Center for Innovation, Invention and Ideas	408-730-7290
Colorado		Denver Public Library	303-640-6220
Delaware		Newark: University of Delaware Library	302-831-2965
District of Columbia		Washington: Founders Library, Howard University	202-806-7252

State	Name of Library	Telephone #
Florida	* Fort Lauderdale: Broward County Main Library	954-357-7444
	* Miami-Dade Public Library	305-375-2665
	Orlando: University of Central Florida Libraries	407-823-2562
	Tampa Campus Library, University of South Florida	813-974-2726
Georgia	Atlanta: Library and Information Center, Georgia Institute of Technology	404-894-4508
Hawaii	* Honolulu: Hawaii State Library	808-586-3477
Idaho	Moscow: University of Idaho Library	208-885-6235
Illinois	Chicago Public Library	312-747-4450
	Springfield: Illinois State Library	217-782-5659
Indiana	Indianapolis-Marion County Public Library	317-269-1741
	West Lafayette: Siegesmund Engineering Library, Purdue University	317-494-2872
Iowa	Des Moines: State Library of Iowa	515-281-4118
Kansas	* Wichita: Ablah Library, Wichita State University	316-978-3155
Kentucky	* Louisville Free Public Library	502-574-1611
Louisiana	Baton Rouge: Troy H. Middleton Library, Louisiana State University	504-388-8875
Maine	Orono: Raymond H. Fogler Library, University of Maine	207-581-1678
Maryland	College Park: Engineering & Physical Sciences Library, University of Maryland	301-405-9157
Massachusetts	Amherst: Physical Sciences and Engineering Library, University of Massachusetts	413-545-1370
	* Boston Public Library	617-536-5400 Ext. 265

State	Name of Library	Telephone #
Michigan	Ann Arbor: Media Union Library, The University of Michigan	313-647-5735
	Big Rapids: Abigail S. Timme Library, Ferris State University	616-592-3602
	** Detroit: Great Lakes Patent and Trademark Center, Detroit Public Library	313-833-3379
Minnesota	* Minneapolis Public Library and Information Center	612-372-6570
Mississippi	Jackson: Mississippi Library Commission	601-359-1036
Missouri	* Kansas City: Linda Hall Library	816-363-4600
	St. Louis Public Library	314-241-2288 Ext. 390
Montana	Butte: Montana Tech of the University of Montana Library	406-496-4281
Nebraska	* Lincoln: Engineering Library, University of Nebraska-Lincoln	402-472-3411
Nevada	Reno: University Library, University of Nevada–Reno	702-784-6500 Ext. 257
New Hampshire	Concord: New Hampshire State Library	603-271-2239
New Jersey	Newark Public Library	201-733-7782
	Piscataway: Library of Science and Medicine, Rutgers University	908-445-2895
New Mexico	Albuquerque: Centennial Science and Engineering Library, The University of New Mexico	505-277-4412
New York	Albany: New York State Library	518-474-5355
	* Buffalo and Erie County Public Library	716-858-7101
	New York: Science, Industry and Business Library, New York Public Library	212-592-7000

State	Name of Library	Telephone #
North Carolina*	Raleigh: D.H. Hill Library, North Carolina State University	919-515-3280
North Dakota	Grand Forks: Chester Fritz Library, University of North Dakota	701-777-4888
Ohio	Akron-Summit County Public Library	330-643-9075
	Cincinnati: The Public Library of Cincinnati and Hamilton County	513-369-6936
	* Cleveland Public Library	216-623-2870
	Columbus: Ohio State University Libraries	614-292-6175
	* Toledo/Lucas County Public Library	419-259-5212
Oklahoma	* Stillwater: Oklahoma State University	405-744-7086
Oregon	* Portland: Lewis & Clark College	503-768-6786
Pennsylvania	* Philadelphia: The Free Library of Philadelphia	215-686-5331
	Pittsburgh: The Carnegie Library of Pittsburgh	412-622-3138
	University Park: Pattee Library, Pennsylvania State University	814-865-4861
Puerto Rico	Mayagüez: General Library, University of Puerto Rico	787-832-4040 Ext. 3459
Rhode Island	Providence Public Library	401-455-8027
South Carolina	Clemson: R.M. Cooper Library, Clemson University	864-656-3024
South Dakota	Rapid City: Devereaux Library, South Dakota School of Mines & Technology	605-394-1275
Tennessee	Memphis & Shelby County Public Library and Information Center	901-725-8877
	Nashville: Stevenson Science and Engineering Library, Vanderbilt University	615-322-2717

State	Name of Library	Telephone #
Texas	Austin: McKinney Engineering Library, The University of Texas at Austin	512-495-4500
	* College Station: Sterling C. Evans Library, Texas A&M University	409-845-3826
	* Dallas Public Library	214-670-1468
	Houston: The Fondren Library, Rice University	713-527-8101 Ext. 2587
	Lubbock: Texas Tech University Library	806-742-2282
Utah	* Salt Lake City: Marriott Library, University of Utah	801-581-8394
Vermont	Burlington: Bailey Howe Library, University of Vermont	802-656-2542
Virginia	* Richmond: James Branch Cabell Library, Virginia Commonwealth University	804-828-1104
Washington	* Seattle: Engineering Library, University of Washington	206-543-0740
West Virginia	Morgantown: Evansdale Library, West Virginia University	304-293-2510 Ext. 113
Wisconsin	Madison: Kurt F. Wendt Library, University of Wisconsin-Madison	608-262-6845
	Milwaukee Public Library	414-286-3051
Wyoming	Casper: Natrona County Public Library	307-237-4935

Note: * Denotes APS-Text access; ** denotes partnership PTDL. Partnership Libraries have access to APS-Text, APS-Image, and X-Search.

Attorneys and Agents

The preparation of an application for patent and the conducting of the proceedings in the Patent and Trademark Office to obtain the patent is an undertaking requiring the knowledge of patent law and rules and Patent and Trademark Office practice and procedures, as well as knowledge of the scientific or technical matters involved in the particular invention.

Inventors may prepare their own applications and file them in the Patent and Trademark Office and conduct the proceedings themselves, but unless they are familiar with these matters or study them in detail, they may get into considerable difficulty. While a patent may be obtained in many cases by persons not skilled in this work, there would be no assurance that the patent obtained would adequately protect the particular invention.

Most inventors employ the services of registered patent attorneys or patent agents. The law gives the Patent and Trademark Office the power to make rules and regulations governing conduct and the recognition of patent attorneys and agents to practice before the Patent and Trademark Office. Persons who are not recognized by the Patent and Trademark Office for this practice are not permitted by law to represent inventors before the Patent and Trademark Office.

The Patent and Trademark Office maintains a register of attorneys and agents. To be admitted to this register, a person must comply with the regulations prescribed by the Office, which require a showing that the person is of good moral character and of good repute and that he/she has the legal, and scientific and technical qualifications necessary to render applicants for patents a valuable service. Certain of these qualifications must be demonstrated by the passing of an examination. Those admitted to the examination must have a college degree in engineering or physical science or the equivalent of such a degree.

The Patent and Trademark Office registers both attorneys at law and persons who are not attorneys at law. The former persons are now referred to as "patent attorneys" and the latter persons are referred to as "patent agents." Insofar as the work of preparing an application for a patent and conducting the prosecution in the Patent and Trademark Office is concerned, patent agents are usually just as well qualified as patent attorneys, although patent agents cannot conduct patent litigation in the courts or perform various services which the local jurisdiction considers as practicing law. For example, a patent agent could not draw up a contract relating to a patent, such as an assignment or a license, if the state in which he/she resides considers drafting contracts as practicing law.

Some individuals and organizations that are not registered advertise their services in the fields of patent searching and invention marketing and development. Such individuals and organizations cannot represent inventors before the Patent and Trademark Office. They are not subject to Patent and Trademark Office discipline, and the Office cannot assist inventors in dealing with them.

The Patent and Trademark Office cannot recommend any particular attorney or agent, or aid in the selection of an attorney or agent, as by stating, in response to inquiry that a named patent attorney, agent, or firm, is "reliable" or "capable." The Patent and Trademark Office publishes a directory of all registered patent attorneys and agents who have indicated their availability to accept new clients, arranged by states, cities, and foreign countries. The Directory may be purchased in paper form from the Government Printing Office or from the PTO's Office of Electronic Information on the Cassis ASSIST CD-ROM disc. It is also available on the PTO Web site.

The telephone directories of most large cities have, in the classified section, a heading for patent attorneys under which those in that area are listed. Many large cities have associations of patent attorneys.

In employing a patent attorney or agent, the inventor executes a power of attorney or authorization of agent which must be filed in the

Patent and Trademark Office and is usually a part of the application papers. When an attorney or agent has been appointed, the Office does not communicate with the inventor directly but conducts the correspondence with the attorney or agent since he/she is acting for the inventor thereafter although the inventor is free to contact the Patent and Trademark Office concerning the status of his/her application. The inventor may remove the attorney or agent by revoking the power of attorney or authorization of agent.

The Patent and Trademark Office has the power to disbar, or suspend from practicing before it, persons guilty of gross misconduct, etc., but this can only be done after a full hearing with the presentation of clear and convincing evidence concerning the misconduct. The Patent and Trademark Office will receive and, in appropriate cases, act upon complaints against attorneys and agents. The fees charged to inventors by patent attorneys and agents for their professional services are not subject to regulation by the Patent and Trademark Office. Definite evidence of overcharging may afford basis for Patent and Trademark Office action, but the Office rarely intervenes in disputes concerning fees.

Disclosure Document

One of the services provided for inventors is the acceptance and preservation by the Patent and Trademark Office for a two year period of papers signed by the inventor(s) disclosing an invention. This disclosure is accepted as evidence of the dates of conception of the invention. The Disclosure Document will be retained for two years and then be destroyed unless it is referred to in a separate letter in a related patent application within those two years.

Disclosure Documents may also be filed at selected Patent and Trademark Depository Libraries (PTDLs), presently including the Sunnyvale (CA) Center for Innovation, Invention and Ideas and Great Lakes Patent and Trademark Center at the Detroit (MI) Public Library. A listing of PTDLs is included in this pamphlet under the heading "Library, Search Room Searches and Patent and Trademark Depository Libraries." One copy of the document is kept at the PTDL and the original documents are sent to the Patent and Trademark Office for fee collection, processing and retention. Disclosure Documents are kept in confidence by the PTO.

The Disclosure Document is not a patent application and the date of its receipt in the PTO does not become the effective filing date of any patent application subsequently filed. The benefits provided by the document will depend upon the adequacy of the disclosure and therefore, it is recommended that the Disclosure Document be a clear and complete explanation of the manner and process of making and using the invention. When the nature of the invention permits, a drawing or sketch should be included.

A fee must accompany the disclosure. See the current fee schedule. To facilitate the PTO's electronic data capture and storage of the Disclosure Document, it must be on white paper having dimensions not to exceed 81/2 x 11 inches (21.6 X 28.0 cm) with each page numbered. Text and drawings must be sufficiently dark to permit reproduction with commonly used office copying machines. Oversized papers, even if foldable to the above dimensions will not be accepted.

Attachments such as videotapes and working models will not be accepted and will be returned.

The Disclosure Document must be accompanied by a separate signed cover letter stating that it is submitted by, or on behalf of, the inventor(s) and requesting that the material be received into the Disclosure Document Program. The original submission will not be returned. A notice with an identifying number and date of receipt in the PTO will be mailed to the customer, indicating that the Disclosure Document may be relied upon only as evidence and that a patent application should be diligently filed if patent protection is desired.

A brochure on Disclosure Documents is available by calling the PTO General Information Services at 1-800-786-9199 or 703-308-4357.

Who May Apply for a Patent

According to the law, only the inventor may apply for a patent, with certain exceptions. If a person who is not the inventor should apply for a patent, the patent, if it were obtained, would be invalid. The person applying in such a case who falsely states that he/she is the inventor would also be subject to criminal penalties. If the inventor is dead, the application may be made by legal representatives, that is, the administrator or executor of the estate. If the inventor is insane, the application for patent may be made by a guardian. If an inventor refuses to apply for a patent or cannot be found, a joint inventor or a person having a proprietary interest in the invention may apply on behalf of the non-signing inventor.

If two or more persons make an invention jointly, they apply for a patent as joint inventors. A person who makes a financial contribution is not a joint inventor and cannot be joined in the application as an inventor. It is possible to correct an innocent mistake in erroneously omitting an inventor or in erroneously naming a person as an inventor.

Officers and employees of the Patent and Trademark Office are prohibited by law from applying for a patent or acquiring, directly or indirectly, except by inheritance or bequest, any patent or any right or interest in any patent.

Application For Patent

Non-Provisional Application for a Patent

A non-provisional application for a patent is made to the Assistant Commissioner for Patents and includes:

(1) A written document which comprises a specification (description and claims), and an oath or declaration;

(2) A drawing in those cases in which a drawing is necessary; and

(3) The filing fee. See the fee schedule.

All application papers must be in the English language or accompanied by a verified translation into the English language along with the required fee set forth in 37 CFR 1.17(k). All application papers must be legibly written either by a typewriter or mechanical printer in permanent dark ink or its equivalent in portrait orientation on flexible, strong, smooth, non-shiny, durable and white paper.

The papers must be presented in a form having sufficient clarity and contrast between the paper and the writing to permit electronic reproduction. The application papers must all be the same size—either 21.0 cm by 29.7 cm (DIN size A4) or 21.6 cm by 27.9 cm (8 1/2 by 11 inches), with a top margins of at least 2.0 cm (3/4 inch), a left side margin of at least 2.5 cm (1 inch), a right side margin of at least 2.0 cm (3/4 inch) and a bottom margin of at least 2.0 cm (3/4 inch) with no holes made in the submitted papers. It is also required that the spacing on all papers be 1 1/2 or double spaced and the application papers must be numbered consecutively (centrally located above or below the text) starting with page one.

The application for patent is not forwarded for examination until all required parts, complying with the rules related thereto, are received. If any application is filed without all the required parts for obtaining a filing date (incomplete or defective), the applicant will be notified of the deficiencies and given a time period to complete the application filing (a surcharge may be required)—at which time a filing date as of the date of such a completed submission will be obtained by the applicant. If the omission is not corrected within a set, specified time period, the application will be returned or otherwise disposed of; the filing fee if submitted will be refunded less a handling fee as set forth in the fee schedule.

It is desirable that all parts of the complete application be deposited in the Office together; otherwise each part must be signed and a letter must accompany each part, accurately and clearly connecting it with the other parts of the application.

All applications received in the PTO are numbered in serial order and the applicant will be informed of the application serial number and filing date by a filing receipt.

The filing date of an application for patent is the date on which the names of the inventors, a specification (including claims) and any required drawings are received in the PTO; or the date on which the last part completing the application are received in the case of a previously incomplete or defective application.

Provisional Application for a Patent

Since June 8, 1995, the PTO has offered inventors the option of filing a provisional application for patent which was designed to provide a lower cost first patent filing in the United States and to give U.S. applicants parity with foreign applicants. Claims and oath or declaration are NOT required for a provisional application. Provisional application provides the means to establish an early effective filing date in a patent application and permits the term "Patent Pending" to be applied in connection with the invention. Provisional applications may not be filed for design inventions.

The filing date of a provisional application is the date on which a written description of the invention, drawings if necessary, and the name of the inventor(s) are received in the PTO. To be complete, a provisional application must also include the filing fee, and a cover sheet specifying that the application is a provisional application for patent. Applicant would then have up to twelve months to file a non-provisional application for patent as described above. The claimed subject matter in the later filed non-provisional application is entitled to the benefit of the filing date of the provisional application if it has support in the provisional application.

Provisional applications are not examined on their merits. A provisional application will become abandoned by the operation of law twelve months from its filing date. The twelve month pendency for a provisional application is not counted toward the 20 year term of a patent granted on a subsequently filed non-provisional application which relies on the filing date of the provisional application.

A surcharge is required for filing the basic filing fee or the cover sheet on a date later than the filing of the provisional application.

A brochure on Provisional Application for Patent is available by calling the PTO General Information Services at 1-800-786-9199 or 703-308-4357 or by accessing PTO's Web site at *http://www.uspto.gov*.

Oath or Declaration, Signature

The oath or declaration of the applicant is required by law for a non-provisional application. The inventor must make an oath or declaration that he/she believes himself/herself to be the original and first inventor of the subject matter of the application, and he/she must make various other allegations required by law and various allegations required by the Patent and Trademark Office rules. The oath must be sworn to by the inventor before a notary public or other officer authorized to administer oaths. A declaration may be used in lieu of an oath as part of the original application for a patent involving designs, plants, and other patentable inventions; for reissue patents; when claiming matter originally shown or described but not originally claimed; or when filing a divisional or continuing application. A declaration does not need to be notarized.

The oath or declaration must be signed by the inventor in person, or by the person entitled by law to make application on the inventor's behalf. A full first and last name with middle initial or name, if any, of each inventor are required. The post office address and citizenship of each inventor are also required.

Sample forms are available by calling the PTO General Information Services at 800-786-9199 or 703-308-4357 or by accessing PTO Web site at *http://www.uspto.gov* under the section titled "PTO Forms."

The papers in a complete application will not be returned for any purpose whatsoever, nor will the filing fee be returned. If applicants have not preserved copies of the papers, the Office will furnish copies for a fee.

Filing Fees*

The filing fee of a nonprovisional application, except in design and plant cases, consists of a basic fee and additional fees. The basic fee entitles the applicant to present twenty (20) claims, including not more than three (3) in independent form. An additional fee is required for each claim in independent form which is in excess of three (3) and an additional fee is required for each claim (whether independent or dependent) which is in excess of a total of twenty (20) claims. If the application contains multiple dependent claims, additional fees are required.

If the owner of the invention is a small entity, (an independent inventor, a small business concern or a non-profit organization), the filing fees are reduced by half if the small entity files a verified statement claiming small entity status.

To avoid errors in the payment of fees it is suggested that the table in the enclosed patent application transmittal letter be utilized to calculate the fee payment.

In calculating fees, a claim is in singularly dependent form if it incorporates by reference a single preceding claim which may be an independent or a dependent claim. A multiple dependent claim or any claim depending therefrom shall be considered as separate dependent claims in accordance with the number of claims to which reference is made.

The law also provides for the payment of additional fees on presentation of additional claims after the application is filed. When an amendment is filed which presents additional claims over the total number already paid for, or additional independent claims over the number of independent claims already accounted for, it must be accompanied by any additional fees due.

* Please Note: The fees are current as of the revision date. Fees are subject to change in October of each year and should therefore be

verified before submission to the PTO. A fee schedule may be obtained by writing to Commissioner of Patents and Trademarks, Washington, D.C. 20231 – Attention: General Information Services or by calling the General Information Services at 800-786-9199 or 703-308-4357 or from the PTO Web site at *http://www.uspto.gov*.

Specification (Description and Claims)

The following order of arrangement should be observed in framing the application:

(a) Application transmittal form.

(b) Fee transmittal form.

(c) Title of the Invention.

(d) Cross Reference to related applications (if any).

(e) Statement of federally sponsored research/development (if any).

(f) Reference to a microfiche appendix (if any).

(g) Background of the Invention.

(h) Brief Summary of the Invention.

(i) Brief description of the several views of the drawing (if any).

(j) Detailed Description of the Invention.

(k) Claim or claims.

(l) Abstract of the disclosure.

(m) Drawings (if any).

(n) Executed oath or declaration.

(o) Sequence listing (if any).

(p) Plant Color Coding Sheet (applicable in plant patent applications).

The specification must include a written description of the invention and of the manner and process of making and using it, and is required to be in such full, clear, concise, and exact terms as to enable any person skilled in the technological area to which the invention pertains, or with which it is most nearly connected, to make and use the same.

The specification must set forth the precise invention for which a patent is solicited, in such manner as to distinguish it from other inventions and from what is old. It must describe completely a specific embodiment of the process, machine, manufacture, composition of matter, or improvement invented, and must explain the mode of operation or principle whenever applicable. The best mode contemplated by the inventor for carrying out the invention must be set forth.

In the case of an improvement, the specification must particularly point out the part or parts of the process, machine, manufacture, or composition of matter to which the improvement relates, and the description should be confined to the specific improvement and to such parts as necessarily cooperate with it or as may be necessary to a complete understanding or description of it.

The title of the invention, which should be as short and specific as possible, should appear as a heading on the first page of the specification, if it does not otherwise appear at the beginning of the application.

A brief abstract of the technical disclosure in the specification including that which is new in the art to which the invention pertains, must be set forth on a separate page immediately following the claims. The abstract should be in the form of a single paragraph of 250 words or less, under the heading "Abstract of the Disclosure."

A brief summary of the invention indicating its nature and substance, which may include a statement of the object of the invention should precede the detailed description. The summary should be commensurate with the invention as claimed and any object recited should be that of the invention as claimed.

When there are drawings, there shall be a brief description of the several views of the drawings, and the detailed description of the invention shall refer to the different views by specifying the numbers of the figures, and to the different parts by use of reference numerals.

The specification must conclude with a claim or claims particularly pointing out and distinctly claiming the subject matter which the applicant regards as the invention. The portion of the application in which the applicant sets forth the claim or claims is an important part of the application, as it is the claims that define the scope of the protection afforded by the patent and which questions of infringement are judged by the courts.

More than one claim may be presented provided they differ substantially from each other and are not unduly multiplied. One or more claims may be presented in dependent form, referring back to and further limiting another claim or claims in the same application. Any dependent claim which refers back to more than one other claim is considered a "multiple dependent claim."

Multiple dependent claims shall refer to such other claims in the alternative only. A multiple dependent claim shall not serve as a basis for any other multiple dependent claim. Claims in dependent form shall be construed to include all of the limitations of the claim incorporated by reference into the dependent claim. A multiple dependent claim shall be construed to incorporate all the limitations of each of the particular claims in relation to which it is being considered.

The claim or claims must conform to the invention as set forth in the remainder of the specification and the terms and phrases used in the claims must find clear support or antecedent basis in the description so that the meaning of the terms in the claims may be ascertainable by reference to the description.

Drawing

The applicant for a patent will be required by law to furnish a drawing of the invention whenever the nature of the case requires a drawing to understand the invention. However, the Commissioner may require a drawing where the nature of the subject matter admits of it; this drawing must be filed with the application. This includes practically all inventions except compositions of matter or processes, but a drawing may also be useful in the case of many processes.

The drawing must show every feature of the invention specified in the claims, and is required by the Office rules to be in a particular form. The Office specifies the size of the sheet on which the drawing is made, the type of paper, the margins, and other details relating to the making of the drawing. The reason for specifying the standards in detail is that the drawings are printed and published in a uniform style when the patent issues, and the drawings must also be such that they can be readily understood by persons using the patent descriptions.

No names or other identification will be permitted within the "sight" of the drawing, and applicants are expected to use the space above and between the hole locations to identify each sheet of drawings. This identification may consist of the attorney's name and docket number or the inventor's name and case number and may include the sheet number and the total number of sheets filed (for example, "sheet 2 of 4"). The following rule, reproduced from title 37 of the Code of Federal Regulations, relates to the standards for drawings:

§ 1.84 Standards for drawings.

(a) Drawings. There are two acceptable categories for presenting drawings in utility patent applications:

(1) Black ink. Black and white drawings are normally required. India ink, or its equivalent that secures solid black lines, must be used for drawings, or

(2) Color. On rare occasions, color drawings may be necessary as the only practical medium by which to disclose the subject matter sought to be patented in a utility patent application or the subject matter of a statutory invention registration. The Patent and Trademark Office will accept color drawings in utility patent applications and statutory invention registrations only after granting a petition filed under this paragraph explaining why the color drawings are necessary. Any such petition must include the following:

(i) The appropriate fee set forth in 37 CFR 1.17(h);
(ii) Three (3) sets of color drawings; and
(iii) The specification must contain the following language as the first paragraph in that portion of the specification relating to the brief description of the drawing:

"*The file of this patent contains at least one drawing executed in color. Copies of this patent with color drawing(s) will be provided by the Patent and Trademark Office upon request and payment of the necessary fee.*"

If the language is not in the specification, a proposed amendment to insert the language must accompany the petition.

(b) Photographs.

(1) Black and white. Photographs are not ordinarily permitted in utility and design patent applications. However, the Office will accept photographs in utility and design patent applications only after granting a petition filed under this paragraph which requests that photographs be accepted. Any such petition must include the following:

(i) The appropriate fee set forth in 37 CFR 1.17(h); and
(ii) Three (3) sets of photographs.

Photographs must either be developed on double weight photographic paper or be permanently mounted on Bristol board. The photographs must be of sufficient quality so that all details in the drawing are reproducible in the printed patent.

(2) Color. Color photographs will be accepted in utility patent applications if the conditions for accepting color drawings have been satisfied. See paragraph (a)(2) of this section.

(c) Identification of drawings. Identifying indicia, if provided, should include the application number or the title of the invention, inventor's name, docket number (if any), and the name and telephone number of a person to call if the Office is unable to match the drawings to the proper application. This information should be placed on the back of each sheet of drawings a minimum distance of 1.5 cm. (5/8 inch) down from the top of the page. In addition, a reference to the application number, or, if an application number has not been assigned, the inventor's name, may be included in the left-hand corner provided that the reference appears within 1.5cm. (9/16 inch) from the top of the sheet.

(d) Graphic forms in drawings. Chemical or mathematical formulae, tables, and waveforms may be submitted as drawings and are subject to the same requirements as drawings. Each chemical or mathematical formula must be labeled as a separate figure, using brackets when necessary, to show that information is properly integrated. Each group of waveforms must be presented as a single figure, using a common vertical axis with time extending along the horizontal axis. Each individual waveform discussed in the specification must be identified with a separate letter designation adjacent to the vertical axis.

(e) **Type of paper**. Drawings submitted to the Office must be made on paper which is flexible, strong, white, smooth, nonshiny, and durable. All sheets must be free from cracks, creases, and folds. Only one side of the sheet shall be used for the drawing. Each sheet must be reasonably free from erasures and must be free from alterations, overwritings, and interlineations. Photographs must either be developed on double weight photographic paper or be permanently mounted on Bristol board. See paragraph (b) of this section for other requirements for photographs.

(f) **Size of paper.** All drawing sheets in an application must be the same size. One of the shorter sides of the sheet is regarded as its top. The size of the sheets on which drawings are made must be:

(1) 21.0 cm. by 29.7 cm. (DIN size A4), or
(2) 21.6 cm. by 27.9 cm. (8 1/2 by 11 inches).

(g) **Margins.** The sheets must not contain frames around the sight; i.e., the usable surface, but should have scan target points, i.e., cross-hairs, printed on two catercorner margin corners. Each sheet must include a top margin of at least 2.5 cm. (1inch), a left side margin of at least 2.5 cm. (1 inch), a right side margin of at least 1.5 cm. (9/16 inch), and a bottom margin of at least 1.0 cm. (3/8 inch), thereby leaving a sight no greater than 17.0 cm. by 26.2 cm. on 21.0 cm. by 29.7 cm. (DIN size A4) drawing sheets, and a sight no greater than 17.6 cm. by 24.4 cm. (6 15/16 by 9 5/8 inches) on 21.6 cm. by 27.9 cm. (8 1/2 by 11 inch) drawing sheets.

(h) **Views.** The drawing must contain as many views as necessary to show the invention. The views may be plan, elevation, section, or perspective views. Detail views of portions of elements, on a larger scale if necessary, may also be used. All views of the drawing must be grouped together and arranged on the sheet(s) without wasting space, preferably in an upright position, clearly separated from one another, and must not be included in the sheets containing the specifications, claims, or abstract. Views must not be connected by projection lines and must not contain center lines. Waveforms of electrical signals may be connected by dashed lines to show the relative timing of the waveforms.

(1) Exploded views. Exploded views, with the separated parts embraced by a bracket, to show the relationship or order of assembly of various parts are permissible. When an exploded view is shown in a figure which is on the same sheet as another figure, the exploded view should be placed in brackets.

(2) Partial views. When necessary, a view of a large machine or device in its entirety may be broken into partial views on a single sheet, or extended over several sheets if there is no loss in facility of understanding the view. Partial views drawn on separate sheets must always be capable of being linked edge to edge so that no partial view contains parts of another partial view. A smaller scale view should be included showing the whole formed by the partial views and indicating the positions of the parts shown. When a portion of a view is enlarged for magnification purposes, the view and the enlarged view must each be labeled as separate views.

(i) Where views on two or more sheets form, in effect, a single complete view, the views on the several sheets must be so arranged that the complete figure can be assembled without concealing any part of any of the views appearing on the various sheets.
(ii) A very long view may be divided into several parts placed one above the other on a single sheet. However, the relationship between the different parts must be clear and unambiguous.

(3) Sectional views. The plane upon which a sectional view is taken should be indicated on the view from which the section is cut by a broken line. The ends of the broken line should be designated by Arabic or Roman numerals corresponding to the view number of the sectional view, and should have arrows to indicate the direction of sight. Hatching must be used to indicate section portions of an object, and must be made by regularly spaced oblique parallel lines spaced sufficiently apart to enable the lines to be distinguished without difficulty. Hatching should not impede the clear reading of the reference characters and lead lines. If it is not possible to place reference characters outside the hatched area, the hatching may be broken off wherever reference characters are inserted. Hatching must be at a substantial angle to the surrounding axes or principal lines, preferably 45°. A cross section must be set out and drawn to show all of the materials as they are shown in the view from which the cross section was taken. The parts

in cross section must show proper material(s) by hatching with regularly spaced parallel oblique strokes, the space between strokes being chosen on the basis of the total area to be hatched. The various parts of a cross section of the same item should be hatched in the same manner and should accurately and graphically indicate the nature of the material(s) that is illustrated in cross section. The hatching of juxtaposed different elements must be angled in a different way. In the case of large areas, hatching may be confined to an edging drawn around the entire inside of the outline of the area to be hatched. Different types of hatching should have different conventional meanings as regards the nature of a material seen in cross section.

(4) <u>Alternate position.</u> A moved position may be shown by a broken line superimposed upon a suitable view if this can be done without crowding; otherwise, a separate view must be used for this purpose.

(5) <u>Modified forms</u>. Modified forms of construction must be shown in separate views.

(i) **Arrangement of views.** One view must not be placed upon another or within the outline of another. All views on the same sheet should stand in the same direction and, if possible, stand so that they can be read with the sheet held in an upright position. If views wider than the width of the sheet are necessary for the clearest illustration of the invention, the sheet may be turned on its side so that the top of the sheet, with the appropriate top margin to be used as the heading space, is on the right-hand side. Words must appear in a horizontal, left-to-right fashion when the page is either upright or turned so that the top becomes the right side, except for graphs utilizing standard scientific convention to denote the axis of abscissas (of X) and the axis of ordinates (of Y).

(j) **View for Official Gazette.** One of the views should be suitable for publication in the Official Gazette as the illustration of the invention.

(k) **Scale.**

(1) The scale to which a drawing is made must be large enough to show the mechanism without crowding when the drawing is reduced in size to two-thirds in reproduction. Views of portions of the mechanism on a larger scale should be used when necessary to show details clearly. Two or more sheets may be used if one does not give sufficient room. The number of sheets should be kept to a minimum.

(2) When approved by the examiner, the scale of the drawing may be graphically represented. Indications such as "actual size" or "scale 1/2" on the drawings, are not permitted, since these lose their meaning with reproduction in a different format.

(3) Elements of the same view must be in proportion to each other, unless a difference in proportion is indispensable for the clarity of the view. Instead of showing elements in different proportion, a supplementary view may be added giving a larger-scale illustration of the element of the initial view. The enlarged element shown in the second view should be surrounded by a finely drawn or "dot-dash" circle in the first view indicating its location without obscuring the view.

(l) **Character of lines, numbers, and letters.** All drawings must be made by a process which will give them satisfactory reproduction characteristics. Every line, number, and letter must be durable, clean, black (except for color drawings), sufficiently dense and dark, and uniformly thick and well-defined. The weight of all lines and letters must be heavy enough to permit adequate reproduction. This requirement applies to all lines however fine, to shading, and to lines representing cut surfaces in sectional views. Lines and strokes of different thicknesses may be used in the same drawing where different thicknesses have a different meaning.

(m) **Shading.** The use of shading in views is encouraged if it aids in understanding the invention and if it does not reduce legibility. Shading is used to indicate the surface or shape of spherical, cylindrical, and conical elements of an object. Flat parts may also be lightly shaded. Such shading is preferred in the case of parts shown in perspective, but not for cross sections. See paragraph (h)(3) of this section. Spaced lines for shading are preferred. These lines must be thin, as few in number as practicable, and they must contrast with the rest of the drawings. As a substitute for shading, heavy lines on the shade side of objects can be used except where they superimpose on each other or obscure reference characters. Light should come from the upper left corner at an angle of 45°. Surface delineations should preferably be shown by proper shading. Solid black shading areas are not permitted, except when used to represent bar graphs or color.

(n) **Symbols.** Graphical drawing symbols may be used for conventional elements when appropriate. The elements for which such symbols and labeled representations are used must be adequately identified in the specification. Known devices should be illustrated by symbols which have a universally recognized conventional meaning and are generally accepted in the art. Other symbols which are not universally recognized may be used, subject to approval by the Office, if they are not likely to be confused with existing conventional symbols, and if they are readily identifiable.

(o) **Legends.** Suitable descriptive legends may be used, or may be required by the Examiner, where necessary for understanding of the drawing, subject to approval by the Office. They should contain as few words as possible.

(p) **Numbers, letters, and reference characters.**

(1) Reference characters (numerals are preferred), sheet numbers, and view numbers must be plain and legible, and must not be used in association with brackets or inverted commas, or enclosed within outlines, e.g., encircled. They must be oriented in the same direction as the view so as to avoid having to rotate the sheet. Reference charac-

ters should be arranged to follow the profile of the object depicted.

(2) The English alphabet must be used for letters, except where another alphabet is customarily used, such as the Greek alphabet to indicate angles, wavelengths, and mathematical formulas.

(3) Numbers, letters, and reference characters must measure at least .32 cm. (1/8 inch) in height. They should not be placed in the drawing so as to interfere with its comprehension. Therefore, they should not cross or mingle with the lines. They should not be placed upon hatched or shaded surfaces. When necessary, such as indicating a surface or cross section, a reference character may be underlined and a blank space may be left in the hatching or shading where the character occurs so that it appears distinct.

(4) The same part of an invention appearing in more than one view of the drawing must always be designated by the same reference character, and the same reference character must never be used to designate different parts.

(5) Reference characters not mentioned in the description shall not appear in the drawings. Reference characters mentioned in the description must appear in the drawings.

(q) Lead lines. Lead lines are those lines between the reference characters and the details referred to. Such lines may be straight or curved and should be as short as possible. They must originate in the immediate proximity of the reference character and extend to the feature indicated. Lead lines must not cross each other. Lead lines are required for each reference character except for those which indicate the surface or cross section on which they are placed. Such a reference character must be underlined to make it clear that a lead line has not been left out by mistake. Lead lines must be executed in the same way as lines in the drawing. See paragraph (l) of this section.

(r) **Arrows.** Arrows may be used at the ends of lines, provided that their meaning is clear, as follows:

(1) On a lead line, a freestanding arrow to indicate the entire section towards which it points;

(2) On a lead line, an arrow touching a line to indicate the surface shown by the line looking along the direction of the arrow; or

(3) To show the direction of movement.

(s) **Copyright or Mask Work Notice.** A copyright or mask work notice may appear in the drawing, but must be placed within the sight of the drawing immediately below the figure representing the copyright or mask work material and be limited to letters having a print size of.32 cm. to .64 cm. (1/8 to 1/4 inches) high. The content of the notice must be limited to only those elements provided for by law. For example, " e1983 John Doe" (17 U.S.C. 401) and "*M* John Doe" (17 U.S.C. 909) would be properly limited and, under current statutes, legally sufficient notices of copyright and mask work, respectively. Inclusion of a copyright or mask work notice will be permitted only if the authorization language set forth in w 1.71(e) is included at the beginning (preferably as the first paragraph) of the specification.

(t) **Numbering of sheets of drawings.** The sheets of drawings should be numbered in consecutive Arabic numerals, starting with 1, within the sight as defined in paragraph (g) of this section. These numbers, if present, must be placed in the middle of the top of the sheet, but not in the margin. The numbers can be placed on the right-hand side if the drawing extends too close to the middle of the top edge of the usable surface. The drawing sheet numbering must be clear and larger than the numbers used as reference characters to avoid confusion. The number of each sheet should be shown by two Arabic numerals placed on either side of an oblique line, with the first being the sheet number and the second being the total number of sheets of drawings, with no other marking.

(u) **Numbering of views.**

(1) The different views must be numbered in consecutive Arabic numerals, starting with 1, independent of the numbering of the sheets and, if possible, in the order in which they appear on the drawing

sheet(s). Partial views intended to form one complete view, on one or several sheets, must be identified by the same number followed by a capital letter. View numbers must be preceded by the abbreviation "FIG." Where only a single view is used in an application to illustrate the claimed invention, it must not be numbered and the abbreviation "FIG." must not appear.

(2) Numbers and letters identifying the views must be simple and clear and must not be used in association with brackets, circles, or inverted commas. The view numbers must be larger than the numbers used for reference characters.

(v) **Security markings.** Authorized security markings may be placed on the drawings provided they are outside the sight, preferably centered in the top margin.

(w) **Corrections.** Any corrections on drawings submitted to the Office must be durable and permanent.

(x) **Holes.** No holes should be made by applicant in the drawing sheets.

The requirements relating to drawings are strictly enforced, but a drawing not complying with all of the regulations may be accepted for purpose of examination, and correction or a new drawing will be required later.

Applicants are advised to employ competent draftsmen to make their drawings.

Models, Exhibits, Specimens

Models are not required in most patent applications since the description of the invention in the specification and the drawings must be sufficiently full, clear, and complete and capable of being understood to disclose the invention without the aid of a model. A model will not be admitted unless specifically requested by the examiner.

A working model, or other physical exhibit, may be required by the Office if deemed necessary. This is not done very often. A working model may be requested in the case of applications for patent for alleged perpetual motion devices.

When the invention relates to a composition of matter, the applicant may be required to furnish specimens of the composition, or of its ingredients or intermediates, for inspection or experiment. If the invention is a microbiological invention, a deposit of the micro-organism involved is required.

Examination of Applications and Proceedings in the Patent and Trademark Office

Applications, other than provisional applications, filed in the Patent and Trademark Office and accepted as complete applications are assigned for examination to the respective examining groups having charge of the areas of technology related to the invention. In the examining group, applications are taken up for examination by the examiner to whom they have been assigned in the order in which they have been filed or in accordance with examining procedures established by the Commissioner.

Applications will not be advanced out of turn for examination or for further action except as provided by the rules, or upon order of the Commissioner to expedite the business of the Office, or upon a verified showing which, in the opinion of the Commissioner, will justify advancing them.

The examination of the application consists of a study of the application for compliance with the legal requirements and a search through United States patents, foreign patent documents, and available literature, to see if the claimed invention is new and nonobvious. A decision is reached by the examiner in the light of the study and the result of the search.

As a result of the examination by the Office, patents are granted in the case of about two out of every three applications for patents which are filed.

Office Action

The applicant is notified in writing of the examiner's decision by an "action" which is normally mailed to the attorney or agent of record. The reasons for any adverse action or any objection or requirement are stated in the action and such information or references are given as may be useful in aiding the applicant to judge the propriety of continuing the prosecution of his/her application.

If the claimed invention is not directed to patentable subject matter, the claims will be rejected. If the examiner finds that the claimed invention lacks novelty or differs only in an obvious manner from what is found in the prior art, the claims may also be rejected. It is not uncommon for some or all of the claims to be rejected on the first office action by the examiner, relatively few applications are allowed as filed.

Applicant's Response

The applicant must request reconsideration in writing, and must distinctly and specifically point out the supposed errors in the examiner's action. The applicant must respond to every ground of objection and rejection in the prior Office action (except that a request may be made that objections or requirements as to form not necessary to further consideration of the claims be held in abeyance until allowable subject matter is indicated). The applicant's action must appear throughout to be a bona fide attempt to advance the case to final action. The mere allegation that the examiner has erred will not be received as a proper reason for such reconsideration.

In amending an application in response to a rejection, the applicant must clearly point out why he/she thinks the amended claims are patentable in view of the state of the art disclosed by the prior references cited or the objections made. He/she must also show how the claims as amended avoid such references or objections.

After response by the applicant the application will be reconsidered, and the applicant will be notified as to the status of the claims, that is, whether the claims are rejected, or objected to, or whether the claims are allowed, in the same manner as after the first examination. The second Office action usually will be made final.

Final Rejection

On the second or later consideration, the rejection or other action may be made final. The applicant's response is then limited to appeal in the case of rejection of any claim and further amendment is restricted. Petition may be taken to the Commissioner in the case of objections or requirements not involved in the rejection of any claim. Response to a final rejection or action must include cancellation of, or appeal from the rejection of, each claim so rejected and, if any claim stands allowed, compliance with any requirement or objection as to form.

In making such final rejection, the examiner repeats or states all grounds of rejection then considered applicable to the claims in the application.

Interviews with examiners may be arranged, but an interview does not remove the necessity for response to Office actions within the required time, and the action of the Office is based solely on the written record.

Restrictions

If two or more inventions are claimed in a single application, and are regarded by the Office to be of such a nature that a single patent should not be issued for both of them, the applicant will be required to limit the application to one of the inventions. The other invention may be made the subject of a separate application which, if filed while the first application is still pending, will be entitled to the benefit of the filing date of the first application. A requirement to restrict the application to one invention may be made before further action by the examiner.

Amendments to Application

Following are some details concerning amendments to the application:

The applicant may amend before or after the first examination and action as specified in the rules, or when and as specifically required by the examiner.

After final rejection or action, amendments may be made canceling claims or complying with any requirement of form which has been made but the admission of any such amendment or its refusal, and any proceedings relative thereto, shall not operate to relieve the application from its condition as subject to appeal or to save it from abandonment.

If amendments touching the merits of the application are presented after final rejection, or after appeal has been taken, or when such amendment might not otherwise be proper, they may be admitted upon a showing of good and sufficient reasons why they are necessary and were not earlier presented.

No amendment can be made as a matter of right in appealed cases. After decision on appeal, amendments can only be made as provided in the rules.

The specifications, claims, and drawing must be amended and revised when required, to correct inaccuracies of description and definition or unnecessary words, and to secure correspondence between the claims, the description, and the drawing.

All amendments of the drawings or specifications, and all additions thereto must not include new matter beyond the original disclosure. Matter not found in either, involving a departure from or an addition to the original disclosure, cannot be added to the application even though supported by a supplemental oath or declaration, and can be shown or claimed only in a separate application.

The claims may be amended by canceling particular claims, by presenting new claims, or by amending the language of particular claims (such amended claims being in effect new claims). In presenting new or amended claims, the applicant must point out how they avoid any reference or ground rejection of record which may be pertinent.

Erasures, additions, insertions, or alterations of the papers and records must not be made by the applicant. Amendments are made by filing a paper, directing or requesting that specified changes or additions be made. The exact word or words to be stricken out or inserted in the application must be specified and the precise point indicated where the deletion or insertion is to be made.

Amendments are "entered" by the Office by making the proposed deletions by drawing a line in red ink through the word or words canceled and by making the proposed substitutions or insertions in red ink. Small insertions are written in at the designated place and larger insertions are indicated by reference.

No change in the drawing may be made except by permission of the Office. Changes in the construction shown in any drawing may be made only by submitting new drawings. A sketch in permanent ink showing proposed changes, to become part of the record, must be filed for approval by the Office before the new drawings are filed. The paper requesting amendments to the drawing should be separate from other papers.

If the number or nature of the amendments render it difficult to consider the case, or to arrange the papers for printing or copying, the examiner may require the entire specification or claims, or any part thereof, to be rewritten.

The original numbering of the claims must be preserved throughout the prosecution. When claims are canceled, the remaining claims must not be renumbered. When claims are added by amendment or substituted for canceled claims, they must be numbered by the applicant consecutively beginning with the number next following the highest numbered claim previously presented. When the application is ready for allowance, the examiner, if necessary, will renumber the claims consecutively in the order in which they appear or in such order as may have been requested by applicant.

Time for Response and Abandonment

The response of an applicant to an action by the Office must be made within a prescribed time limit. The maximum period for response is set at 6 months by the statute which also provides that the Commissioner may shorten the time for reply to not less than 30 days. The usual period for response to an Office action is 3 months. A shortened time for reply may be extended up to the maximum 6-month period. An extension of time fee is normally required to be paid if the response period is extended. The amount of the fee is dependent upon the length of the extension. If no reply is received within the time period, the application is considered as abandoned and no longer pending. However, if it can be shown that the failure to prosecute was unavoidable or unintentional, the application may be revived by the Commissioner. The revival requires a petition to the Commissioner, and a fee for the petition, which should be filed without delay. The proper response must also accompany the petition if it has not yet been filed.

Appeal to the Board of Patent Appeals and Interferences and to the Courts

If the examiner persists in the rejection of any of the claims in an application, or if the rejection has been made final, the applicant may appeal to the Board of Patent Appeals and Interferences in the Patent and Trademark Office. The Board of Patent Appeals and Interferences consists of the Commissioner of Patents and Trademarks, the Deputy Commissioner, the Assistant Commissioners, and the administrative patent judges, but normally each appeal is heard by only three members. An appeal fee is required and the applicant must file a brief to support his/her position. An oral hearing will be held if requested upon payment of the specified fee.

As an alternative to appeal, in situations where an applicant desires consideration of different claims or further evidence, a new continuation application is often filed. The new application requires a filing fee and should submit the claims and evidence for which consideration is desired. If it is filed before expiration of the period for appeal and specific reference is made therein to the earlier application, applicant will be entitled to the earlier filing date for subject matter common to both applications.

If the decision of the Board of Patent Appeals and Interferences is still adverse to the applicant, an appeal may be taken to the Court of Appeals for the Federal Circuit or a civil action may be filed against the Commissioner in the United States District Court for the District of Columbia. The Court of Appeals for the Federal Circuit will review the record made in the Office and may affirm or reverse the Office's action. In a civil action, the applicant may present testimony in the court, and the court will make a decision.

Interferences

Occasionally two or more applications are filed by different inventors claiming substantially the same patentable invention. The patent can only be granted to one of them, and a proceeding known as an "interference" is instituted by the Office to determine who is the first inventor and entitled to the patent. About one percent of the applications filed become involved in an interference proceeding. Interference proceedings may also be instituted between an application and a patent already issued, provided the patent has not been issued for more than one year prior to the filing of the conflicting application, and provided that the conflicting application is not barred from being patentable for some other reason.

Each party to such a proceeding must submit evidence of facts proving when the invention was made. In view of the necessity of proving the various facts and circumstances concerning the making of the invention during an interference, inventors must be able to produce evidence to do this. If no evidence is submitted a party is restricted to the date of filing the application as his/her earliest date. The priority question is determined by a board of three administrative patent judges on the evidence submitted. From the decision of the Board of Patent Appeals and Interferences, the losing party may appeal to the Court of Appeals for the Federal Circuit or file a civil action against the winning party in the appropriate United States district court.

The terms "conception of the invention" and "reduction to practice" are encountered in connection with priority questions. Conception of the invention refers to the completion of the devising of the means for accomplishing the result. Reduction to practice refers to the actual construction of the invention in physical form: in the case of a machine it includes the actual building of the machine, in the case of an article or composition it includes the actual making of the article or composition, in the case of a process it includes the actual carrying out of the steps of the process. Actual operation, demonstration, or testing for the intended use is also usually necessary. The filing of a regular application for patent completely disclosing the invention is treated as equivalent to reduction to practice. The inventor who proves to be the first to conceive the invention and the first to reduce it to practice will be held to be the prior inventor, but more complicated situations cannot be stated this simply.

Allowance and Issue of Patent

If, on examination of the application, or at a later stage during the reconsideration of the application, the patent application is found to be allowable, a notice of allowance will be sent to the applicant, or to applicant's attorney or agent of record, if any, and a fee for issuing the patent is due within three months from the date of the notice. If timely payment of the issue fee is not made, the application will be regarded as abandoned. See current fee schedule.

A provision is made in the statute whereby the Commissioner may accept the fee late, when the delay is shown to be unavoidable. When the issue fee is paid, the patent issues as soon as possible after the date of payment, dependent upon the volume of printing on hand. The patent grant then is delivered or mailed on the day of its grant, or as soon thereafter as possible, to the inventor's attorney or agent if there is one of record, otherwise directly to the inventor. On the date of the grant, the patent file becomes open to the public. Printed copies of the specification and drawing are available on the same date.

In case the publication of an invention by the granting of a patent would be detrimental to the national defense, the patent law gives the Commissioner the power to withhold the grant of the patent and to order the invention kept secret for such period of time as the national interest requires.

Nature of Patent and Patent Rights

The patent is issued in the name of the United States under the seal of the Patent and Trademark Office, and is either signed by the Commissioner of Patents and Trademarks or has his name written thereon and attested by an Office official. The patent contains a grant to the patentee, and a printed copy of the specification and drawing is annexed to the patent and forms a part of it. The grant confers "the right to exclude others from making, using, offering for sale or selling the invention throughout the United States or importing the invention into the United States" and its territories and possessions for which the term of the patent shall be 20 years from the date on which the application for the patent was filed in the United States or , if the application contains a specific reference to an earlier filed application under 35 U.S.C. 120, 121 or 365(c), from the date of the earliest such application was filed, and subject to the payment of maintenance fees as provided by law.

The exact nature of the right conferred must be carefully distinguished, and the key is in the words "right to exclude" in the phrase just quoted. The patent does not grant the right to make, use, offer for sale or sell or import the invention but only grants the exclusive nature of the right. Any person is ordinarily free to make, use, offer for sale or sell or import anything he/she pleases, and a grant from the Government is not necessary. The patent only grants the right to exclude others from making, using, offering for sale or selling or importing the invention. Since the patent does not grant the right to make, use, offer for sale, or sell, or import the invention, the patentee's own right to do so is dependent upon the rights of others and whatever general laws might be applicable. A patentee, merely because he/she has received a patent for an invention, is not thereby authorized to make, use, offer for sale, or sell, or import the invention if doing so would violate any law. An inventor of a new automobile who has obtained a patent thereon would not be entitled to use the patented automobile in violation of the laws of a State requiring a license, nor may a patentee sell an article, the sale of which may be forbidden by a law, merely because a patent has been obtained.

Neither may a patentee make, use, offer for sale, or sell, or import his/her own invention if doing so would infringe the prior rights of others. A patentee may not violate the Federal antitrust laws, such as by resale price agreements or entering into combination in restraints of trade, or the pure food and drug laws, by virtue of having a patent. Ordinarily there is nothing which prohibits a patentee from making, using, offering for sale, or selling, or importing his/her own invention, unless he/she thereby infringes another's patent which is still in force.

The term of the patent shall be 20 years from the date on which the application for the patent was filed in the United States or, if the application contains a specific reference to an earlier filed application under 35 U.S.C. 120, 121 or 365(c), from the date of the earliest such application was filed, and subject to the payment of maintenance fees as provided by law. A maintenance fee is due 3 1/2, 7 1/2 and 11 1/2 years after the original grant for all patents issuing from the applications filed on and after December 12, 1980. The maintenance fee must be paid at the stipulated times to maintain the patent in force. After the patent has expired anyone may make, use, offer for sale or sell or import the invention without permission of the patentee, provided that matter covered by other unexpired patents is not used. The terms may be extended for certain pharmaceuticals and for certain circumstances as provided by law.

Maintenance Fees

All utility patents which issue from applications filed on and after December 12, 1980 are subject to the payment of maintenance fees which must be paid to maintain the patent in force. These fees are due at 3 1/2, 7 1/2 and 11 1/2 years from the date the patent is granted and can be paid without a surcharge during the "window-period" which is the six month period preceding each due date, e.g., 3 years to 3 years and six months. (See fee schedule for a list of maintenance fees.)

Failure to pay the current maintenance fee on time may result in expiration of the patent. A 6-month grace period is provided when the maintenance fee may be paid with a surcharge. The grace period is the 6-month period immediately following the due date. The Patent and Trademark Office does not mail notices to patent owners that maintenance fees are due. If, however, the maintenance fee is not paid on time, efforts are made to remind the responsible party that the maintenance fee may be paid during the grace period with a surcharge.

Patents relating to some pharmaceutical inventions may be extended by the Commissioner for up to five years to compensate for marketing delays due to Federal premarketing regulatory procedures. Patents relating to all other types of inventions may be extended for certain circumstances as provided by law.

Correction of Patents

Once the patent is granted, it is outside the jurisdiction of the Patent and Trademark Office except in a few respects. The Office may issue without charge a certificate correcting a clerical error it has made in the patent when the printed patent does not correspond to the record in the Office. These are mostly corrections of typographical errors made in printing. Some minor errors of a typographical nature made by the applicant may be corrected by a certificate of correction for which a fee is required. The patentee may disclaim one or more claims of his/her patent by filing in the Office a disclaimer as provided by the statute.

When the patent is defective in certain respects, the law provides that the patentee may apply for a reissue patent. This is a patent granted to replace the original and is granted only for the balance of the unexpired term. However, the nature of the changes that can be made by means of the reissue are rather limited; new matter cannot be added.

Any person may file a request for reexamination of a patent, along with the required fee, on the basis of prior art consisting of patents or printed publications. At the conclusion of the reexamination proceedings, a certificate setting forth the results of the reexamination proceeding is issued.

Assignments and Licenses

A patent is personal property and may be sold to others or mortgaged; it may be bequeathed by a will, and it may pass to the heirs of a deceased patentee. The patent law provides for the transfer or sale of a patent, or of an application for patent, by an instrument in writing. Such an instrument is referred to as an assignment and may transfer the entire interest in the patent. The assignee, when the patent is assigned to him or her, becomes the owner of the patent and has the same rights that the original patentee had.

The statute also provides for the assignment of a part interest, that is, a half interest, a fourth interest, etc., in a patent. There may also be a grant which conveys the same character of interest as an assignment but only for a particularly specified part of the United States.

A mortgage of patent property passes ownership thereof to the mortgagee or lender until the mortgage has been satisfied and a retransfer from the mortgagee back to the mortgagor, the borrower, is made. A conditional assignment also passes ownership of the patent and is regarded as absolute until canceled by the parties or by the decree of a competent court.

An assignment, grant, or conveyance of any patent or application for patent should be acknowledged before a notary public or officer authorized to administer oaths or perform notarial acts. The certificate of such acknowledgment constitutes prima facie evidence of the execution of the assignment, grant, or conveyance.

Recording of Assignments

The Office records assignments, grants, and similar instruments sent to it for recording, and the recording serves as notice. If an assignment, grant, or conveyance of a patent or an interest in a patent (or an application for patent) is not recorded in the Office within three months from its date, it is void against a subsequent purchaser for a valuable consideration without notice, unless it is recorded prior to the subsequent purchase.

An instrument relating to a patent should identify the patent by number and date (the name of the inventor and title of the invention as stated in the patent should also be given). An instrument relating to an application should identify the application by its application number and date of filing, the name of the inventor, and title of the invention as stated in the application should also be given. Sometimes an assignment of an application is executed at the same time that the application is prepared and before it has been filed in the Office. Such assignment should adequately identify the application, as by its date of execution and name of the inventor and title of the invention, so that there can be no mistake as to the application intended.

If an application has been assigned and the assignment is recorded, on or before the date the issue fee is paid, the patent will be issued to the assignee as owner. If the assignment is of a part interest only, the patent will be issued to the inventor and assignee as joint owners.

Joint Ownership

Patents may be owned jointly by two or more persons as in the case of a patent granted to joint inventors, or in the case of the assignment of a part interest in a patent. Any joint owner of a patent, no matter how small the part interest, may make, use, offer for sale and sell and import the invention for his or her own profit provided they do not infringe another's patent rights, without regard to the other owners, and may sell the interest or any part of it, or grant licenses to others, without regard to the other joint owner, unless the joint owners have made a contract governing their relation to each other. It is accordingly dangerous to assign a part interest without a definite agreement between the parties as to the extent of their respective rights and their obligations to each other if the above result is to be avoided.

The owner of a patent may grant licenses to others. Since the patentee has the right to exclude others from making, using, offering for sale or selling or importing the invention, no one else may do any of these things without his/her permission. A patent license agreement is in essence nothing more than a promise by the licensor not to sue the licensee. No particular form of license is required; a license is a contract and may include whatever provisions the parties agree upon, including the payment of royalties, etc.

The drawing up of a license agreement (as well as assignments) is within the field of an attorney at law. Such attorney should be familiar with patent matters as well. A few States have prescribed certain formalities to be observed in connection with the sale of patent rights.

Infringement of Patents

Infringement of a patent consists of the unauthorized making, using, offering for sale or selling any patented invention within the United States or United States Territories, or importing into the United States of any patented invention during the term of the patent. If a patent is infringed, the patentee may sue for relief in the appropriate Federal court. The patentee may ask the court for an injunction to prevent the continuation of the infringement and may also ask the court for an award of damages because of the infringement. In such an infringement suit, the defendant may raise the question of the validity of the patent, which is then decided by the court. The defendant may also aver that what is being done does not constitute infringement. Infringement is determined primarily by the language of the claims of the patent and, if what the defendant is making does not fall within the language of any of the claims of the patent, there is no literal infringement.

Suits for infringement of patents follow the rules of procedure of the Federal courts. From the decision of the district court, there is an appeal to the Court of Appeals for the Federal Circuit. The Supreme Court may thereafter take a case by writ of certiorari. If the United States Government infringes a patent, the patentee has a remedy for damages in the United States Court of Federal Claims. The Government may use any patented invention without permission of the patentee, but the patentee is entitled to obtain compensation for the use by or for the Government.

The Office has no jurisdiction over questions relating to infringement of patents. In examining applications for patent, no determination is made as to whether the invention sought to be patented infringes any prior patent. An improvement invention may be patentable, but it might infringe a prior unexpired patent for the invention improved upon, if there is one.

Patent Marking and "Patent Pending"

A patentee who makes or sells patented articles, or a person who does so for or under the patentee is required to mark the articles with the word "Patent" and the number of the patent. The penalty for failure to mark is that the patentee may not recover damages from an infringer unless the infringer was duly notified of the infringement and continued to infringe after the notice.

The marking of an article as patented when it is not in fact patented is against the law and subjects the offender to a penalty. Some persons mark articles sold with the terms "Patent Applied For" or "Patent Pending." These phrases have no legal effect, but only give information that an application for patent has been filed in the Patent and Trademark Office. The protection afforded by a patent does not start until the actual grant of the patent. False use of these phrases or their equivalent is prohibited.

Design Patents

The patent laws provide for the granting of design patents to any person who has invented any new, original and ornamental design for an article of manufacture. The design patent protects only the appearance of an article, and not its structure or utilitarian features. The proceedings relating to granting of design patents are the same as those relating to other patents with a few differences.

See current fee schedule for the filing fee for a design application. A design patent has a term of 14 years from grant, and no fees are necessary to maintain a design patent in force. If on examination it is determined that an applicant is entitled to a design patent under the law, a notice of allowance will be sent to the applicant or applicant's attorney, or agent, calling for the payment of an issue fee.

The drawing of the design patent conforms to the same rules as other drawings, but no reference characters are required.

The specification of a design application is short and ordinarily follows a set form. Only one claim is permitted, following a set form.

Plant Patents

The law also provides for the granting of a patent to anyone who has invented or discovered and asexually reproduced any distinct and new variety of plant, including cultivated sports, mutants, hybrids, and newly found seedlings, other than a tuber-propagated plant or a plant found in an uncultivated state.

Asexually propagated plants are those that are reproduced by means other than from seeds, such as by the rooting of cuttings, by layering, budding, grafting, inarching, etc.

With reference to tuber-propagated plants, for which a plant patent cannot be obtained, the term "tuber" is used in its narrow horticultural sense as meaning a short, thickened portion of an underground branch. Such plants covered by the term "tuber-propagated" are the Irish potato and the Jerusalem artichoke.

An application for a plant patent consists of the same parts as other applications with the addition of a plant color coding sheet. The term of a plant patent shall be 20 years from the date on which the application for the patent was filed in the United States or, if the application contains a specific reference to an earlier filed application under 35 U.S.C. 120, 121 or 365(c), from the date of the earliest such application was filed.

The application papers for a plant patent and any responsive papers pursuant to the prosecution must be filed in duplicate but only one need be signed (in the case of the application papers the original should be signed); the second copy may be a legible copy of the original. The reason for providing an original and duplicate file is that the duplicate file is sent to the Agricultural Research Service, Department of Agriculture for an advisory report on the plant variety.

The specification should include a complete detailed description of the plant and the characteristics thereof that distinguish the same over related known varieties, and its antecedents, expressed in botanical terms in the general form followed in standard botanical text books or publications dealing with the varieties of the kind of plant involved (evergreen tree, dahlia plant, rose plant, apple tree, etc.), rather than a mere broad non-botanical characterization such as commonly found in nursery or seed catalogs. The specification should also include the origin or parentage of the plant variety sought to be patented and must particularly point out where and in what manner the variety of plant has been asexually reproduced. Where color is a distinctive feature of the plant, the color should be positively identified in the specification by reference to a designated color as given by a recognized color dictionary. Where the plant variety originated as a newly found seedling, the specification must fully describe the conditions (cultivation, environment, etc.) under which the seedling was found growing to establish that it was not found in an uncultivated state.

A plant patent is granted on the entire plant. It therefore follows that only one claim is necessary and only one is permitted.

The oath or declaration required of the applicant in addition to the statements required for other applications must include the statement that the applicant has asexually reproduced the new plant variety.

Plant patent drawings are not mechanical drawings and should be artistically and competently executed. The drawing must disclose all the distinctive characteristics of the plant capable of visual representation. When color is a distinguishing characteristic of the new variety, the drawing must be in color. Two duplicate copies of color drawings must be submitted. All color drawings should be so mounted as to provide a 2-inch margin at the top for office markings when the patent is printed.

Specimens of the plant variety, its flower or fruit, should not be submitted unless specifically called for by the examiner.

The filing fee on each plant application and the issue fee can be found in the fee schedule. For a qualifying small entity filing and issue fees are reduced by half.

All inquiries relating to plant patents and pending plant patent applications should be directed to the Patent and Trademark Office and not to the Department of Agriculture.

The Plant Variety Protection Act (Public Law 91-577), approved December 24, 1970, provides for a system of protection for sexually reproduced varieties, for which protection was not previously provided, under the administration of a Plant Variety Protection Office within the Department of Agriculture. Requests for information regarding the protection of sexually reproduced varieties should be addressed to Commissioner, Plant Variety Protection Office, Agricultural Marketing Service, National Agricultural Library Bldg., Room 500, 10301 Baltimore Blvd., Beltsville, Md. 20705-2351.

Treaties and Foreign Patents

Since the rights granted by a United States patent extend only throughout the territory of the United States and have no effect in a foreign country, an inventor who wishes patent protection in other countries must apply for a patent in each of the other countries or in regional patent offices. Almost every country has its own patent law, and a person desiring a patent in a particular country must make an application for patent in that country, in accordance with the requirements of that country.

The laws of many countries differ in various respects from the patent law of the United States. In most foreign countries, publication of the invention before the date of the application will bar the right to a patent. In most foreign countries maintenance fees are required. Most foreign countries require that the patented invention must be manufactured in that country after a certain period, usually three years. If there is no manufacture within this period, the patent may be void in some countries, although in most countries the patent may be subject to the grant of compulsory licenses to any person who may apply for a license.

There is a treaty relating to patents which is adhered to by 140 countries, including the United States, and is known as the Paris Convention for the Protection of Industrial Property. It provides that each country guarantees to the citizens of the other countries the same rights in patent and trademark matters that it gives to its own citizens. The treaty also provides for the right of priority in the case of patents, trademarks and industrial designs (design patents). This right means that, on the basis of a regular first application filed in one of the member countries, the applicant may, within a certain period of time, apply for protection in all the other member countries. These later applications will then be regarded as if they had been filed on the same day as the first application. Thus, these later applicants will have priority over applications for the same invention which may have been filed during the same period of time by other persons. Moreover, these later applications, being based on the first application, will not be invalidated by any acts accomplished in the interval, such as, for example, publication or exploitation of the invention, the sale of copies of the design, or use of the trademark. The period of time mentioned above, within which the subsequent applications may be filed in the other countries, is 12 months in the case of first applications for patent and six months in the case of industrial designs and trademarks.

Another treaty, known as the Patent Cooperation Treaty, was negotiated at a diplomatic conference in Washington, D.C., in June of 1970. The treaty came into force on January 24, 1978, and is presently adhered to by over 90 countries, including the United States. The treaty facilitates the filing of applications for patent on the same invention in member countries by providing, among other things, for centralized filing procedures and a standardized application format.

The timely filing of an international application affords applicants an international filing date in each country which is designated in the international application and provides (1) a search of the invention and (2) a later time period within which the national applications for patent must be filed.

A number of patent attorneys specialize in obtaining patents in foreign countries. In general, an inventor should be satisfied that he/she could make some profit from foreign patents or that there is some particular reason for obtaining them, before he/she attempts to apply for foreign patents.

Under United States law it is necessary, in the case of inventions made in the United States, to obtain a license from the Commissioner of Patents and Trademarks before applying for a patent in a foreign country. Such a license is required if the foreign application is to be filed before an application is filed in the United States or before the expiration of six months from the filing of an application in the United States. The filing of an application for patent constitutes the request for a license and the granting or denial of such request is indicated in the filing receipt mailed to each applicant. After six months from the United States filing, a license is not required unless the invention has been ordered to be kept secret. If the invention has been ordered to be kept secret, the consent to the filing abroad must be obtained from the Commissioner of Patents and Trademarks during the period the order of secrecy is in effect.

Foreign Applicants for United States Patents

The patent laws of the United States make no discrimination with respect to the citizenship of the inventor. Any inventor, regardless of his/her citizenship, may apply for a patent on the same basis as a U.S. citizen. There are, however, a number of particular points of special interest to applicants located in foreign countries.

The application for patent in the United States must be made by the inventor and the inventor must sign the oath or declaration (with certain exceptions), differing from the law in many countries where the signature of the inventor and an oath of inventorship are not necessary. If the inventor is dead, the application may be made by his/her executor or administrator, or equivalent, and in the case of mental disability it may be made by his/her legal representative (guardian).

No United States patent can be obtained if the invention was patented abroad before applying in the United States by the inventor or his/her legal representatives or assigns on an application filed more than 12 months before filing in the United States. Six months are allowed in the case of a design patent.

An application for a patent filed in the United States by any person who has previously regularly filed an application for a patent for the same invention in a foreign country which affords similar privileges to citizens of the United States shall have the same force and effect for the purpose of overcoming intervening acts of others as if filed in the United States on the date on which the application for a patent for the same invention was first filed in such foreign country. This is the case, provided the application in the United States is filed within 12 months (six months in the case of a design patent) from the earliest date on which any such foreign application was filed. A copy of the foreign application certified by the patent office of the country in which it was filed is required to secure this right of priority.

If any application for patent has been filed in any foreign country by the applicant or by his/her legal representatives or assigns prior to his/her application in the United States, the applicant must, in the oath or declaration accompanying the application, state the country in which the earliest such application has been filed, giving the date of filing the application. All applications filed more than a year before the filing in the United States must also be recited in the oath or declaration.

An oath or declaration must be made with respect to every application. When the applicant is in a foreign country the oath or affirmation may be before any diplomatic or consular officer of the United States, or before any officer having an official seal and authorized to administer oaths in the foreign country, whose authority shall be proved by a certificate of a diplomatic or consular officer of the United States. The oath is attested in all cases by the proper official seal of the officer before whom the oath is made.

When the oath is taken before an officer in the country foreign to the United States, all the application papers (except the drawing) must be attached together and a ribbon passed one or more times through all the sheets of the application, and the ends of the ribbons brought together under the seal before the latter is affixed and impressed, or each sheet must be impressed with the official seal of the officer before whom the oath was taken.

If the application is filed by the legal representative (executor, administrator, etc.) of a deceased inventor, the legal representative must make the oath or declaration.

When a declaration is used, the ribboning procedure is not necessary, nor is it necessary to appear before an official in connection with the making of a declaration.

A foreign applicant may be represented by any patent attorney or agent who is registered to practice before the United States Patent and Trademark Office.

Answers to Questions Frequently Asked

1. Q. What do the terms "patent pending" and "patent applied for" mean?

A. They are used by a manufacturer or seller of an article to inform the public that an application for patent on that article is on file in the Patent and Trademark Office. The law imposes a fine on those who use these terms falsely to deceive the public.

2. Q. Is there any danger that the Patent and Trademark Office will give others information contained in my application while it is pending?

A. No. All patent applications are maintained in the strictest secrecy until the patent is issued. After the patent is issued, however, the Office file containing the application and all correspondence leading up to issuance of the patent is made available in the Files Information Room for inspection by anyone and copies of these files may be purchased from the Office.

3. Q. May I write to the Patent and Trademark Office directly about my application after it is filed?

A. The Office will answer an applicant's inquiries as to the status of the application, and inform you whether your application has been rejected, allowed, or is awaiting action. However, if you have a patent attorney or agent of record in the application file the Office will not correspond with both you and the attorney/agent concerning the merits of your application. All comments concerning your application should be forwarded through your attorney or agent.

4. Q. Is it necessary to go to the Patent and Trademark Office to transact business concerning patent matters?

A. No; most business with the Office is conducted by correspondence. Interviews regarding pending applications can be arranged with examiners if necessary, however, and are often helpful.

5. Q. If two or more persons work together to make an invention, to whom will the patent be granted?

A. If each had a share in the ideas forming the invention, they are joint inventors and a patent will be issued to them jointly on the basis of a proper patent application. If, on the other hand, one of these persons has provided all of the ideas of the invention, and the other has only followed instructions in making it, the person who contributed the ideas is the sole inventor and the patent application and patent shall be in his/her name alone.

6. Q. If one person furnishes all of the ideas to make an invention and another employs him or furnishes the money for building and testing the invention, should the patent application be filed by them jointly?

A. No. The application must be signed by the true inventor, and filed in the Patent and Trademark Office, in the inventors name. This is the person who furnishes the ideas, not the employer or the person who furnishes the money.

7. Q. Does the Patent and Trademark Office control the fees charged by patent attorneys and agents for their services?

A. No. This is a matter between you and your patent attorney or agent in which the Office takes no part. To avoid misunderstanding you may wish to ask for estimate charges for: (a) the search (b) preparation of the patent application, (c) Patent and Trademark Office prosecution.

8. Q. Will the Patent and Trademark Office help me to select a patent attorney or agent to make my patent search or to prepare and prosecute my patent application?

A. No. The Office cannot make this choice for you. However, your own friends or general attorney may help you in making a selection from among those listed as registered practitioners on the Office roster. Also, some bar associations operate lawyer referral services that maintain lists of patent lawyers available to accept new clients.

9. Q. Will the Patent and Trademark Office advise me as to whether a certain patent promotion organization is reliable and trustworthy?

A. No. The Office has no control over such organizations and does not supply information about them. It is advisable, however, to check on the reputation of invention promotion firms before making any commitments. It is suggested that you obtain this information by inquiring of the Better Business Bureau of the city in which the organization is located, or of the bureau of commerce and industry or bureau of consumer affairs of the state in which the organization has its place of business. You may also undertake to make sure that you are dealing with reliable people by asking your own patent attorney or agent or by inquiry of others who may know them.

10. Q. Are there any organizations in my area which can tell me how and where I may be able to obtain assistance in developing and marketing my invention?

A. Yes. In your own or neighboring communities you may inquire of such organizations as chambers of commerce, and banks. Many communities have locally financed industrial development organizations which can help you locate manufacturers and individuals who might be interested in promoting your idea.

11. Q. Are there any state government agencies that can help me in developing and marketing of my invention?

A. Yes. In nearly all states there are state planning and development agencies or departments of commerce and industry which seek new product and new process ideas to assist manufacturers and communities in the state. If you do not know the names or addresses of your state organizations you can obtain this information by writing to the governor of your state.

12. Q. Can the Patent and Trademark Office assist me in the developing and marketing of my patent?

A. The Office cannot act or advise concerning the business transactions or arrangements that are involved in the development and marketing of an invention. However, the Office will publish, at the request of a patent owner, a notice in the Official Gazette that the patent is available for licensing or sale. The fee for this is $25.

☆ U. S. GOVERNMENT PRINTING OFFICE: 1997-417-391/82901